THE NUMBER
&
THE FORCE

πGπGπGπGπGπGπ

PI & GRAVITATION

Merle A. Barlow
Calvary Christian High School

WESTBOW®
PRESS
A DIVISION OF THOMAS NELSON
& ZONDERVAN

Scripture taken from the King James Version of the Bible.

WestBow Press books may be ordered through booksellers or by contacting:

WestBow Press
A Division of Thomas Nelson & Zondervan
1663 Liberty Drive
Bloomington, IN 47403
www.westbowpress.com
1 (866) 928-1240

ISBN: 978-1-4908-5322-2 (sc)
ISBN: 978-1-4908-5323-9 (hc)
ISBN: 978-1-4908-5324-6 (e)

Library of Congress Control Number: 2014917174

Printed in the United States of America.

WestBow Press rev. date: 10/08/2014

This book is dedicated to my Mother, ANN MARIE BARLOW. The reason for this tribute is not because my Mom was significantly interested in Science or Mathematics. She was an intelligent, compassionate woman who loved and supported her family. As I developed the information about the two subjects of my book, mathematical pi and the law of gravitation, I thought of a specific kind of different pi – apple pie. The apple, because of Isaac Newton's observation, is a historical symbol of gravitation. It was a logical transition from these thoughts to the proverbial expression "Mom and Apple Pie". This phrase is intended to connote sentiments regarding concepts that are good and pleasant. That could not be more appropriate as a description for my Mother. Her culinary skill (including apple pie) was one of numerous talents that blessed me and my Dad. My Dad provided the leadership and protection for our family, and my Mom was the heart of our home.

Note to Reader

This book is a brief analysis of two concepts. There is no intended continuity between these two separate and independent subjects, except of course, that I desired to include both of them in the same book.

Part 1 Mathematical Pi (π)

Part 2 Gravitation

CONTENTS
PART 1
Pi

C O N T E N T S

PART 2
Gravitation

CONTENTS

PART 1

MATHEMATICAL PI

Frontispiece

A View of the First Digit of π

(Reference Backispiece for a View of the Last Digit of π)

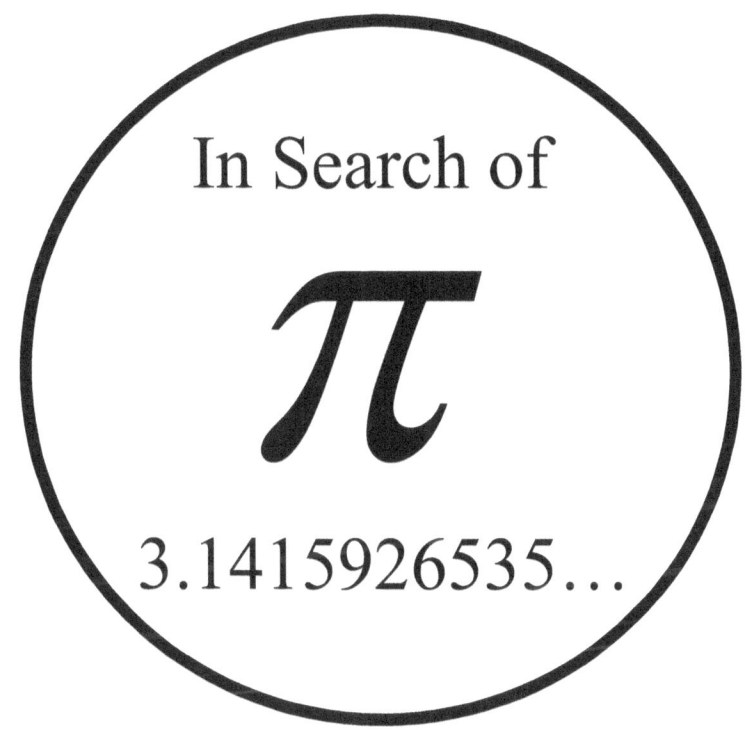

In Search of

π

3.1415926535…

There is probably no symbol, and the number that it represents, that has provoked as much interest, confusion, and mystery as the number pi (π).

✱ Caveat ✱

Information Regarding Calculation Accuracy
Limitations and Display Limitations

A TI-84 Plus Graphic Calculator was used for all calculations in this document. Because of the intrinsic limitations of this particular instrument, register overflows occurred during the accumulations of repeated iterations for the calculation of increasingly more accurate approximations of π. Therefore, although some of the formulas and iterative equations can theoretically provide an infinite number of decimal digits for the approximations of π, the calculator's internal registers cannot accommodate the large values of the interim variables, or the large number of digits calculated for the approximations. Consequently, regardless of the number of iterations or the number of computed decimal digits, the maximum number of total digits displayed is ten, and the maximum number of decimal digits is nine. Further accuracy is beyond the range of this calculator.

Preface

My original intent for this article was to present just the principle of exhaustion method for calculating approximations of the value of pi (Reference Chapters 4 & 5). This technique was prevalent in the early history of our world. Antiphon and Bryson, the scholarly contemporaries of Socrates, were first to articulate the exhaustion principle about 430 B.C. About 200 years later, Archimedes (one of the greatest thinkers in history) used the same principle during the 3rd century. However, unlike Antiphon and Bryson who used areas of polygons to approximate pi, Archimedes focused on the perimeter of the polygons to approximate the circle's circumference. The principle of exhaustion method, used by many scholars for more than 2000 years, began to lose its appeal during the early 1600s because of the advancement in mathematics that provided more efficient methods of approximating pi.

It is profoundly humbling to recognize that Archimedes' accuracy was accomplished without the benefit of a symbol for zero or without decimal notation. These concepts had not yet been introduced and would not be for hundreds of years to come. The achievements of scholars involved in scientific analysis prior to the 20th century were notable considering that they did not have the advantages of desk calculators (late 1940s) or computers (first in 1949). Today, of course, we have the capabilities of our handheld calculators, personal computers, and magnificent supercomputers.

I enlarged the scope of my presentation to include several examples of methods for estimating pi that were subsequent to the principle of exhaustion efforts. These additional techniques will provide a more complete understanding of the progress that has occurred. Although our quantitative and qualitative understanding of pi has significantly increased, the discovery of the true richness of the nature of this number may be many years away.

Clearwater, Florida
November, 2014

Merle

Prologue

It is not difficult to construct a circle. With a little more effort, one can construct a square. The circle can be thought of as nature's perfect form – it exists naturally everywhere by God's design, and is associated with the concept of infinity. Conversely, the square can be thought of as man's perfect form, and is associated with the concept of finiteness. The square is prevalent by man's design, but rarely occurs naturally.

In early civilization, it was discovered that there was a special relationship between the circumference and the diameter of a circle. The ratio of the circle's circumference to its diameter is represented by the symbol π (pi). (π is the sixteenth letter of the Greek alphabet, and is equivalent to the lower case English alphabetic "p"). It was discovered that this value could not be expressed as the ratio of two integers. As more and more accurate approximations of this ratio were achieved, it was discovered that there was not a repeating pattern of digits.

More puzzling, however, was the discovery that by using Euclidean theorems (compass and straightedge), it is impossible to construct a square that has the same area as a given circle. This is referred to as "squaring the circle" (it is also called the quadrature of the circle). It has been <u>proven</u> that this procedure is impossible, but that has not prevented many people who continue the futile effort. [Note: The ancient Greeks defined the two conditions for the circle squaring problem. In addition to using only compass and straightedge, the solution must be performed without using an infinite number of steps. Of course, if either of these two conditions is disregarded, it is a simple matter to square the circle. Using advanced mathematics, like calculus, a square equal in area to a given circle can be constructed quite easily].

For thousands of years it seemed natural and reasonable that a mathematical or geometrical relationship would be discovered between the circle and the square. It seemed logical that the circle could be measured as squares and rectangles are measured. That, however, is incorrect. And, our puzzling friend, the value called pi, is responsible for this perplexing situation. We continue to explore and examine additional digits of pi with the anticipation of discovering some clue to the complete and complex nature of this mysterious number.

1
Pictures of Pi

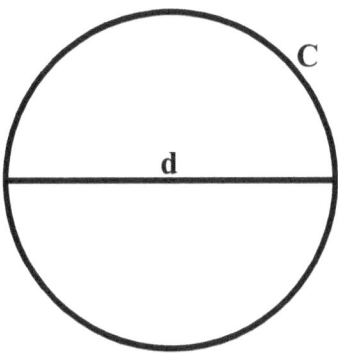

Figure 1

In Figure 1, the ratio of the circumference (C)
of a circle to its diameter (d) is pi.

$$\frac{C}{d} = \pi$$

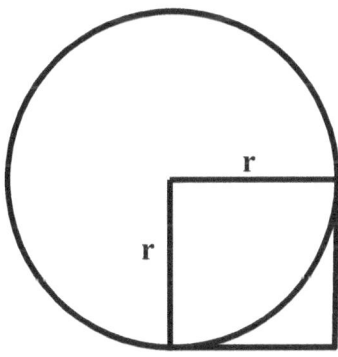

Figure 2

In Figure 2, the ratio of the area of the circle
to the area of the square is pi.

$$\text{Area of circle} = \pi r^2$$

$$\text{Area of square} = r^2$$

$$\frac{\pi r^2}{r^2} = \pi$$

Pictures of Pi

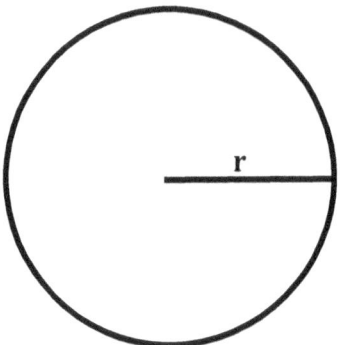

Figure 3

In Figure 3, if the radius of the circle equals 1, the area of the circle is pi.

$$A = \pi r^2$$
$$A = \pi(1)^2$$
$$A = \pi$$

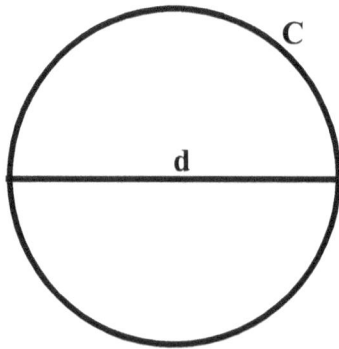

Figure 3a

In Figure 3a, if the diameter of the circle equals 1, the circumference is pi.

$$C = \pi d$$
$$C = \pi(1)$$
$$C = \pi$$

2
A Piece of Pi

Actually, it could be argued that the piece of pie illustrated below is specifically $\dfrac{\pi}{4}$.

[O is the center of the circle.]

(Hold this page up to a mirror).

3
The Great Pyramid and Pi

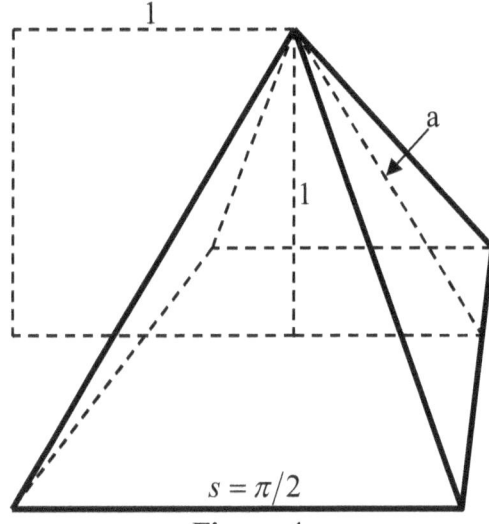

Figure 4

Let height (altitude) = 1
Then:
Each side of base (s) = $\pi/2$
Altitude of lateral face (a) =
$$\frac{\sqrt{\pi^2+16}}{4}$$
One-half of a base side = $\pi/4$
Area of lateral face = 1
(approximately)
Area of square = 1

The Great Pyramid at Giza in northern Egypt near Cairo has an interesting relationship inherent in its structure. The ratio of the length of one side of the base to the height is approximately $\pi/2$. The historian Herodotus wrote that the pyramid was constructed so that the area of each lateral face would equal the area of a square whose side is equal to the pyramid's height. It can be demonstrated that any pyramid constructed with this condition will necessarily approximate pi. In Figure 4, given that the ratio of the length of one side of the base to the height is equal to $\pi/2$, and when we let the height of the pyramid equal 1, then each side of the base will equal $\pi/2$.

The inherent relationship: $s/h = \pi/2$

If $h = 1$, $s = \pi/2$

Determining the area of a lateral face:

$$(1)^2 + (\pi/4)^2 = a^2$$

$$\frac{\pi^2}{16} + 1 = a^2$$

$$\frac{\pi^2 + 16}{16} = a^2$$

$$1/4\sqrt{\pi^2 + 16} = a$$

Since the area of the lateral face (a triangle) is equal to one-half the product of its base $(\pi/2)$ and altitude $\left(\dfrac{\sqrt{\pi^2+16}}{4}\right)$, we have

$$(1/2)(\pi/2)\left(\frac{\sqrt{\pi^2+16}}{4}\right) = \frac{\pi\sqrt{\pi^2+16}}{16} = .999 \text{ (approximately 1)}$$

Regardless of the value for the height or side, if the ratio of the side to the height is equal to $\pi/2$, the area of the square and lateral face will be equal. Conversely, if the area of the lateral face is equal to the area of the square (whose side is equal to the pyramid's height), the ratio of the side to the height will be $\pi/2$.

4
Principle of Exhaustion Introduction

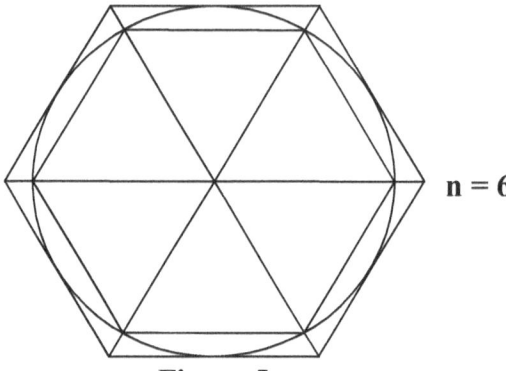

n = 6

Figure 5

A circle with inscribed and circumscribed polygons illustrates the principle of exhaustion method of measuring the value of π. This method was used by many scholars, but probably the clearest description of this method was given by Archimedes. Essentially, it is a limiting process in which a unit circle (radius = 1) is squeezed between two regular polygons – one inscribed in the circle, and one circumscribed about the circle. In Figure 5, the circumference or area of the circle is approximated by a 6-sided inscribed polygon and a 6-sided circumscribed polygon. For each value of n, the perimeter or area of the inscribed polygon is less than the circumference or area of the circle, and the perimeter or area of the circumscribed polygon is greater than the circumference or area of the circle. As n increases, the perimeters and areas become better and better approximations of the circumference and area of the circle. For example, if you begin with a hexagon, then double its sides, and then double them again, and continue to double the sides, eventually there will be so many sides that both the inscribed and circumscribed polygons will be virtually equal to the circle (perimeter and area).The lower boundary (inscribed polygon) approaches the circle as a limit as n becomes infinitely large. The upper boundary (circumscribed polygon) approaches the circle as a limit as n becomes infinitely large. The circumference and area of the circle are between these two boundaries. As n increases, so does the accuracy of the approximations for the circumference and area of the circle, and likewise, the accuracy (number of digits) of the approximation of π also increases.

5
Principle of Exhaustion Details

Estimation of the Value of π for a Unit Circle (radius =1) that has Inscribed and Circumscribed Regular Polygons with an Increasing Number of Sides (Principle of Exhaustion).

Detail for Inscribed Polygons

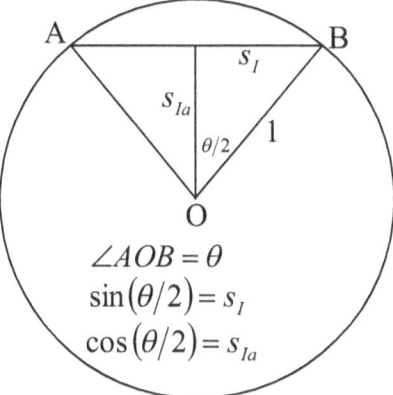

$$\angle AOB = \theta$$
$$\sin(\theta/2) = s_I$$
$$\cos(\theta/2) = s_{Ia}$$

Detail for Circumscribed Polygons

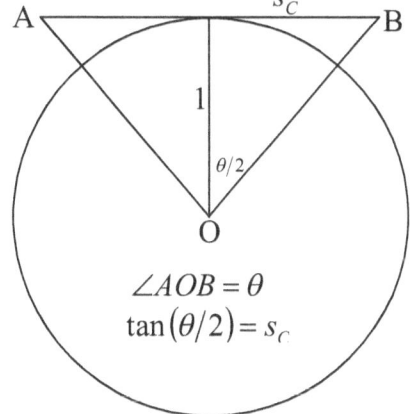

$$\angle AOB = \theta$$
$$\tan(\theta/2) = s_C$$

Figure 6

In Figure 6, the central angle AOB (θ) represents one of n central angles of the n-sided polygons, and is the same angle for both the inscribed and the circumscribed polygons. Separate drawings are illustrated so that the details can be more easily presented.

Definitions and Equations for the Variables in Table 1

Note: For all polygon approximations of π:
(1) The lower boundary value of the polygon perimeter is not equal to the lower boundary value of the polygon area.
(2) The upper boundary value of the polygon perimeter is equal to the upper boundary value of the polygon area.
n = number of polygon sides

θ = a central angle of the n-sided polygon $\left(\dfrac{360°}{n}\right)$

$\dfrac{\theta}{2}$ = one-half of the central angle $\left(\dfrac{360°}{2n}\right)$

$\sin\left(\dfrac{\theta}{2}\right) = s_I = \dfrac{inscribed\ polygon\ side}{2}$

$(s_{Ia})(s_I)$ = area of inscribed central angle (θ)

$\tan\left(\dfrac{\theta}{2}\right) = s_C = \dfrac{circumscribed\ polygon\ side}{2}$

s_C = area of circumscribed central angle (θ)

$C_I = (2)(s_I)(n)$ = perimeter of inscribed polygon

$C_C = (2)(s_C)(n)$ = perimeter of circumscribed polygon

$C = 2\pi r$ = perimeter (circumference) of circle bounded by the inscribed and circumscribed polygons

$\pi = \dfrac{C}{2r}$, where C = circumference, and r = 1

Therefore, the lower and upper boundaries of the estimate for π is calculated by dividing the respective polygon perimeters by 2:

$$\dfrac{C_I}{2} < \pi < \dfrac{C_C}{2}$$

$\dfrac{C_I}{2}$ represents the lower boundary of π for a polygon of n sides

$\dfrac{C_C}{2}$ represents the upper boundary of π for a polygon of n sides

$A_I = (s_{Ia})(s_I)(n)$ = area of inscribed polygon

$A_C = (s_C)(n)$ = area of circumscribed polygon

$A = \pi r^2$ = area of circle bounded by the inscribed and circumscribed polygons

$\pi = A$, where A = circle area, and r = 1

Therefore, the lower and upper boundaries of the estimate for π is determined by A_I and A_C :

$$A_I < \pi < A_C$$

A_I represents the lower boundary of π for a polygon of n sides

A_C represents the upper boundary of π for a polygon of n sides

Table 1 – Principle of Exhaustion (Perimeter Method)

The accuracy of the calculated values in this table through n = 196,608 were unaffected by any calculator limitations.

Table 1

No. of Sides n	Central Angle $360°/n = \theta$	Lower Boundary of π $C_I/2$	Upper Boundary of π $C_C/2$	Good Digits	Remark
6	60	3	3.464101615	1	
12	30	3.105828541	3.215390309		
24	15	3.132628613	3.159659942	2	
48	7.5	3.139350203	3.146086215		
96	3.75	3.141031951	3.1427146	3	Ref. [1]
192	1.875	3.141452472	3.14187305	4	
384	.9375	3.141557608	3.141662747		
768	.46875	3.141583892	3.141610177		
1,536	.234375	3.141590463	3.141597034	6	Practical Use Digits
3,072	.1171875	3.141592106	3.141593749		
6,144	.05859375	3.141592517	3.141592927	7	
12,288	.029296875	3.141592619	3.141592722		
24,576	.0146484375	3.141592645	3.141592692	8	Ref. [2]
49,152	.0073242188	3.141592651	3.141592679		
98,304	.0036621094	3.141592653	3.141592676		
196,608	.0018310547	3.141592653	3.141592654	9	Valid Max. Display
393,216	.0009155273438	3.141592654	3.141592654	(10)	Ref. [3] & [4]
786,432	.0004577636719	3.141592654	3.141592654		
1,572,864	.0002288818359	3.141592654	3.141592654		$n = (6)(2^{18})$

References for Table 1

[1] 3rd Century B.C.:
Using 96-sided polygons, the great mathematician Archimedes calculated the lower boundary as 3.14084507 and the upper boundary as 3.142857143. Note that these two values are very close to the correct values calculated in my table. The first 3 digits are accurate – the same accuracy I calculated for n = 96. Note also that Archimedes' two boundary values are the rational values $3\frac{10}{71}$ and $3\frac{1}{7}$.

[2] 450 A.D.:
Using 24,576-sided polygons, the father and son astronomer team of Tsu Ch'ung-chih and Tsu Keng-chih calculated the values of π as approximately $355/113$, or about 3.1415929 (which is only 8-millionths of 1 percent different from the true value of 3.14159265…). Notice that the first 7 digits of the astronomers' calculation are correct. My calculation for n = 24,576 yields 8 accurate digits.

[3] 1579 A.D.:
Using 393,216-sided polygons, a French lawyer and amateur mathematician named Viete determined that π was greater than 3.1415926535 and less than 3.1415926537. At the time, this was the most precise measurement of pi in history. Notice that the first 10 digits of the lower boundary are accurate (3.141592653…).This is the same accuracy that I have calculated, except that my tenth digit is correctly rounded up to 4 (display limitation).

[4] For n = 393,216 and greater, the calculator display limitation prevents the viewing of accurate digits beyond the eighth decimal digit.

Consider the significance of the achievements of the people mentioned above, as well as many others who did not have the advantages of advanced mathematics and technologies to develop their analysis. It is indeed a simple task for me, using my calculator, to examine and verify their efforts.

Now consider a mathematician named Ludolf van Ceulen who spent many years calculating pi using the same method that Archimedes and many others had used (inscribed and circumscribed polygons). His effort was a monumental feat of patience and endurance. At the time of his death in 1610, van Ceulen had manually calculated 35 digits of pi that involved polygons of more than 32 billion sides – yes, more than 32,000,000,000!

The Exhaustion method's laborious operations involving polygons began to lose its prominence in the 1600s. Either better methods would have to be discovered, or the reality of knowing only a relatively few digits would have to be accepted. As mathematics progressed, more effective methods were found to study and search for pi.

6
John Wallis' Formula

$$\frac{\pi}{2} = \frac{2 \times 2 \times 4 \times 4 \times 6 \times 6 \times 8 ...}{1 \times 3 \times 3 \times 5 \times 5 \times 7 \times 7 ...}$$

The Englishman John Wallis was a mathematician and cryptographer. The above equation was discovered by Wallis in 1655, and is an infinite product that involves only rational operations with no awkward roots. This product slowly converges on pi – the first term is higher (4), the second term is lower (2.67), the third term is again higher (3.56), etc. The product of the seven terms illustrated above is 3.34.

7
Gregory-Leibniz Series

$$\arctan x = x - \frac{x^3}{3} + \frac{x^5}{5} - \frac{x^7}{7} + \frac{x^9}{9} - \frac{x^{11}}{11} \dots$$

There were many great mathematicians living in the seventeenth century. A few of the notable ones were Pascal, Kepler, Fermat, Gregory, Leibniz, and Isaac Newton. Each of these prominent scholars made significant contributions in the quest for understanding the nature of pi, and in the discovery and development of calculus.

In the early 1670s, Gregory and Leibniz independently discovered an elegant solution to calculate arctangents which led to a new way of calculating pi. The reason for this was that the tangent of $45°$ is 1. $45°$ equals $\frac{\pi}{4}$ in radians. Therefore, the arctangent of 1 is equivalent to $\frac{\pi}{4}$ radians. By substituting 1 for x in the series, $\frac{\pi}{4}$ is easily approximated.

$$\frac{\pi}{4} = 1 - \frac{1}{3} + \frac{1}{5} - \frac{1}{7} + \frac{1}{9} - \frac{1}{11} \dots$$

While this series is impressive in its elegance and simplicity, and for what it reveals about the nature of pi, it is painfully inefficient when it comes to actually calculating digits. It would require 300 terms of this series to obtain just 2 decimal digits of pi, and thousands more to become practically useful.

8
Sir Isaac Newton

$$\frac{\pi}{6} = \frac{1}{2} + \frac{1}{2}\left(\frac{1}{3\times 2^3}\right) + \frac{1\times 3}{2\times 4}\left(\frac{1}{5\times 2^5}\right) + \frac{1\times 3\times 5}{2\times 4\times 6}\left(\frac{1}{7\times 2^7}\right) + \ldots$$

Leibniz and Newton independently discovered calculus, and are recognized as cofounders. Newton's brilliance is renowned, but his involvement with pi is not (actually, this effort was a relatively simple task for him, but one of relatively low priority considering his abilities). Finding the ratio of a circle's circumference to its diameter was no longer just a question of basic calculations. The development of the calculus and the arctangent series enabled mathematicians to make faster calculations than measuring polygons.

About 1665, Newton had found several infinite series that approximated pi. He determined at least the first 16 decimal digits. Calculating just four terms of one of Newton's series given above produces the approximation 3.141.

Although Newton did not play as significant a role as others in the analysis of pi, excessive tribute for this exceptional genius is impossible. He was a nonpareil!

9
Euler's Formulas

(1) $\dfrac{\pi}{4} = 2\arctan\left(\dfrac{1}{3}\right) + \arctan\left(\dfrac{1}{7}\right)$

(2) $\dfrac{\pi}{4} = 5\arctan\left(\dfrac{1}{7}\right) + 2\arctan\left(\dfrac{3}{79}\right)$

(3) $\arctan x = \displaystyle\sum_{n=0}^{\infty}\left[\dfrac{2^{2n}(n!)^2}{(2n+1)!} \times \dfrac{x^{2n+1}}{(1+x^2)^{n+1}}\right]$

Euler, one of the greatest and most prolific mathematicians, made significant contributions to the pi effort during the middle of the eighteenth century. He found many arctangent formulas and infinite series that approximated pi.

Both equations (1) and (2) estimate pi accurately to at least ten digits (at least nine decimal digits) to yield:

3.14159265(4)

Obviously, my calculator rounds the ninth decimal digit up to 4. So, considering the internal calculator limitation and the limit of a maximum of ten display digits, I do not know how many accurate digits are produced by these two equations.

In equation (3):

Substituting $\dfrac{\sqrt{3}}{3}$ for x in arctan x, $\arctan\left(\dfrac{\sqrt{3}}{3}\right) = \dfrac{\pi}{6}$

The values in Table 2 indicate the approximation for the specified values of n. Likewise, other values of x, with their corresponding angles, could be chosen for other approximations.

For $x = \dfrac{\sqrt{3}}{3}$, this infinite series approximates $\dfrac{\pi}{6}$.

Table 2

n	Approximation of π	Accurate Digits
0	2.598076211	
1	3.031088913	1
2	3.117691454	2
3	3.136249141	
4	3.140373071	3
5	3.141310328	4
6	3.141526618	5
7	3.141577086	
8	3.141588961	
9	3.141591773	6
10	3.141592443	7
11	3.141592603	8
12	3.141592641	
13	3.141592651	9
14	3.141592653	10
15	3.141592653	
16	3.14159265(4)	

For n > 14, this equation is of course producing additional digits of π – they are just not viewable with my calculator.
At n = 35, there was an overflow error.

10
Ramanujan Equation

$$\frac{1}{\pi} = \frac{2\sqrt{2}}{9801} \sum_{n=0}^{\infty} \left[\frac{(4n)}{(n!)^4} \times \frac{1103 + 26390n}{(4 \times 99)^{4n}} \right]$$

In the early 1900s, an Indian mathematician named Ramanujan brought amazing insights to the study of pi. Scientists and mathematicians continue to study the equations developed by this genius, using them to generate other algorithms that are efficiently used with computers. The equation above is an iterative equation. The result of each iteration is added to the previous results to provide more and more accurate approximations of pi. Some iterations will provide significantly more relevant digits than previous iterations.

As Table 3 illustrates, this equation is a powerful algorithm that quickly produces many significant digits of pi. Note that in just the second iteration (n=1), at least the first 10 digits of π are found.

Table 3

n	Approximation of π	Accurate Digits
0	3.14159273	6
1	3.14159265(4)	10

For n = 1, the ninth decimal digit is apparently rounded up to 4. For n = 2 through 9, the approximations are the same, and additional digit accuracy is not discernible with my calculator. For n = 10, there is an overflow error.

11
Chudnovsky Brothers Equation

$$\frac{1}{\pi} = 12 \sum_{0}^{\infty} \left[(-1)^n \times \frac{(6n)!}{(n!)^3 (3n)!} \times \frac{13591409 + 545140134n}{640320^{(3n+3/2)}} \right]$$

Probably no one has pursued the study of the nature of pi with the dedication of the two Chudnovsky brothers, world-class mathematicians and number theorists. Together, the brothers have held several world records for calculating the largest number of digits of pi. The first record was 480 million digits, then one billion, then 2 billion. In 1996, they held the record at over 8 billion digits.

They have developed extremely sophisticated equations to describe pi. Some people may think that calculating more and more digits is a frivolous activity, but they do not understand that the quest is to find the appearance of some rules that will distinguish the digits of pi from other numbers, to bring us closer to an understanding of the mathematics and physics of the universe in which we were created. The digits of pi appear to be so random that if there were a rule to the sequence, it may require trillions of digits to begin to discern it.

Calculating a record-breaking number of digits for pi also challenges any computer, exposing any flaw in hardware or software where a problem may occur once in a billion calculations. Finding these logic errors or design oversights is not uncommon when you push a supercomputer at over 100 billion calculations every minute continually for days and weeks. Computing pi is like the ultimate stress test for a computer. These computer problems must be found and resolved, because a problem that causes one specific digit to be incorrect will probably also cause subsequent digits to be inaccurate. Extraordinary understanding, dedication, and effort have earned the Chudnovsky brothers a permanent place in the history of pi.

The very powerful and sophisticated equation illustrated above is one of the equations developed by the Chudnovsky brothers. This particular equation is an iterative, infinite series equation that efficiently approximates pi. Table 4 illustrates the results of performing this equation with my calculator.

Table 4

n	Approximation of π	Accurate Digits
0	3.14159265(4)	10
1	3.14159265(4)	10

Note that the very first iteration (n = 0) produced at least 10 accurate digits of pi. (The ninth decimal digit is apparently rounded up to 4).

For iterations 1 through 5, the result is the same – no additional digits are discernible because of the calculator limitation.

Iteration 6 produced an overflow condition – again, because of the calculator limitation.

12
Pi Time

Many people are aware of the day that is logically chosen to celebrate pi. That day is March 14 (3/14). At 1:59 PM (or AM) on that day, the designation for month, day, and time represent the first six digits of pi (3.14159).

I propose that there is a more interesting and paradoxical perspective of pi in regard to time. Consider the pi approximation 3.141592653589... Remember that this is only an approximation, because the number of digits of pi is infinite (does not end). This approximation represents a specific time expressed as a decimal fraction. In other words, what clock time does this approximation of pi represent? Using the 12 decimal digits illustrated above, let's convert that decimal portion of pi to clock time:

60 minutes x .141592653589 = 8.495559215 minutes

This is 8.495559215 minutes after 3 o'clock. Since we have .495559215 of another minute, we can convert this fraction of a minute to seconds:

60 seconds x .495559215 = 29.73355292 seconds

So now we have 8 minutes, and 29.73355292 seconds after 3 o'clock. Rounding to the nearest tenth of a second:

pi time is 3 hours, 8 minutes, and 29.7 seconds

However, this time is less than the actual value of pi. It is also true that 3 hours, 8 minutes, and 29.8 seconds would be a time greater than pi. So, the actual pi time will occur some time between 3 hours, 8 minutes, 29.7 seconds, and 3 hours, 8 minutes, 29.8 seconds – or will it?

The pi time will occur when the clock time is <u>exactly</u> <u>equal</u> to the value of pi. That time will have to occur when the clock time is greater than 29.7 seconds and less than 29.8 seconds. As the clock time progresses beyond 29.7 seconds to closer and closer approach the value of pi, can it ever actually equal that value? As the clock time increases, there will continue to be more and more digits (decimal places) of pi that the clock time must equal in value. But we know that eventually, 29.8 seconds will occur, and that time is greater than the value of pi.

So, how does the clock time attain the value of pi between 29.7 and 29.8 seconds? It does not! What? It does not!

This is not a figment of fantasy or fallacious fiction. I conclude this matter with the following explanation: Assume the clock time is 3 hours, 8 minutes, and exactly 29.7 seconds. This time is less than the pi value time. .03 of a second later, it will be 29.73 seconds and closer to the pi value time. Another .005 of a second later, the clock time is now 29.735 – getting closer to pi.

You can imagine the progression through the subsequent digits – thousands of digits, millions, billions without end. Yes, the clock time will continually get closer and closer to that pi value time, but can never equal it because the clock time cannot equate to the last digit of pi. It cannot, because there is no last digit. Trying to equate a time to a value that is infinite is a paradoxical effort.

13
Pi Records

π

In 1997, Kanada and Takahashi calculated over 51 billion digits of pi on a Hitachi SR2201 computer in just over 29 hours.

π

In September of 1999, Professor Kanada of the University of Tokyo calculated 206,158,430,000 decimal digits of pi (approximately (3×2^{36}).

π

In September of 2002, Kanada and his team broke their own world record, calculating 1.2411 trillion digits (1,241,100,000,000) which was more than six times the previous record. The new record was accomplished in over 400 hours with a Hitachi supercomputer capable of 2 trillion calculations per second. Kanada's team spent five years designing the program that was used to calculate pi and test the efficiency of the supercomputer.

π

In late 2009, Fabrice Bellard, a computer scientist achieved a record of nearly 2.7 trillion digits of pi.

π

In August of 2010, Shigeru Kondo, a Japanese systems engineer, and Alexander J. Yee, an American computer science student, calculated the value of pi to 5 trillion digits in 90 days.

π

The current world record of 10 trillion digits was accomplished by Kondo and Yee in October of 2011, doubling their previous record set in 2010.

14
Elusive Equality

The arrows represent a Roman Numeral equation that is invalid.
(2 does <u>not</u> equal 22 divided by eight).
Move <u>**one**</u> arrow to make the equation approximately true.

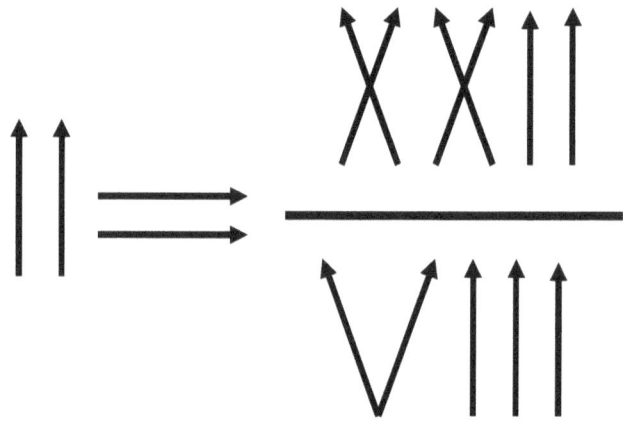

[Reference the Appendix for the answer].

15
A Few Friendly Factoids

(1)
The first zero in the value of π occurs at the 32nd decimal digit.

(2)
The millionth digit of π is 1.

(3)
The sequence "123456789" first appears at the $523,551,502^{nd}$ decimal digit.

(4)
If just a billion digits of π were printed in ordinary type, the expression would extend over 1,200 miles.

(5)
For the mathematically inclined, another definition of π:
Twice some number between 0 and 2 whose cosine is 0.

(6)
There are various techniques for memorizing π. The most common method is the word-length mnemonic in which the number of letters in each word equals a digit of pi. A simple mnemonic is:
"How I wish I could calculate pi" (3.141592).
Personally, I prefer a Spanish mnemonic:
"Sol y Luna y Mundo proclaman al Eterno Autor del Cosmo".
(Sun and Moon and World acclaim the Eternal Author of the Cosmos).
Not only does it praise the Creator of all Reality, it provides more digits (3.1415926535).

(7)
There is confusion in some circles about the nature of π – some say that pi are square – others insist that pi are round. Reference Figure 3, and Figure 3a of Chapter 1, "Pictures of Pi" for the answer.

(8)
Ten decimal digits of pi are sufficient to calculate the circumference of the earth to a fraction of an inch; 39 digits are

sufficient to calculate the circumference of a circle the size of the observable universe with an error no greater than the radius of a hydrogen atom.

(9)
I (the author) personally know someone who knows **all** the digit of pi. ☺
Do you know this Someone?
        [Reference the Appendix for the answer].
(10)
Oh, perhaps if you are wondering, the 10 trillionth digit of pi is 5.

(11)
What is the result of dividing the circumference of a jack-o'-lantern by its diameter?
        [Reference the Appendix for the answer].

(12)
Table 5 illustrates the number of occurrences of the digits 0 through 9 in the first million decimal places of pi.

Table 5

Digit	Number of Occurrences
0	99,959
1	99,758
2	100,026
3	100,229
4	100,230
5	100,359
6	99,548
7	99,800
8	99,985
9	100,106

16
Squaring the Circle

The history of mankind is replete with the efforts to change something into something different. Notable among these efforts, alchemists attempted to change one element into another, e.g., lead into gold. This was one of many attempts through alchemy, magic, and science to be successful at transformation.

In the Prologue, I described the transformation of the area of a circle into a square with an equal area. Several additional comments regarding this Euclidian transformation are appropriate. This process of attempting to square the circle (constructing a square that has the same area as a given circle) has been proven to be impossible. It is true that the proof provided in 1761 that pi is irrational (it cannot be expressed as a ratio of two integers) discouraged some circle squarers, but this fact was not convincing, because it is possible to construct some irrational numbers geometrically (e.g., $\sqrt{2}$). However, the impossibility was firmly established in 1882 when the number pi was proven to <u>also</u> be a transcendental number. A transcendental number cannot be described by a finite algebraic equation. In other words, pi cannot be the root of an algebraic equation with rational coefficients, and therefore it cannot possibly be constructed within Euclidian Geometry constraints. This fact, regrettably, did not and will not dissuade or discourage many who continue their hopeless efforts to square the circle.

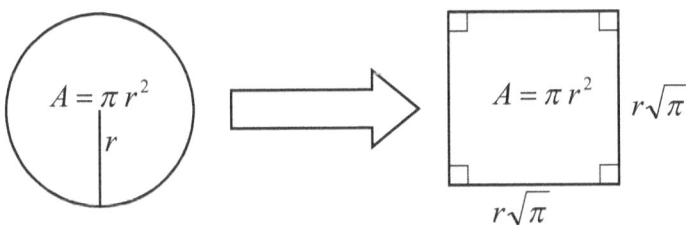

If you know both the circle's diameter and circumference, you know π. The mathematicians of antiquity were unaware that it was not possible to precisely know the ratio of a circle's circumference to its diameter, and consequently all the subsequent attempts to square the circle were futile.

17
Searching Pi

Here is something fun to do – find your birthday in Pi. There is a computer program that gives you the capability to enter any string of digits (up to 120 digits) to search for an exact match in the first 200 million digits of pi. You can search for your birthday, or any other sequence of digits you desire. To accomplish this, simply Search the Web for **"www.angio.net"**, then select the **"Pi-Search Page"** option, and then enter your birthday digits at the **"Search For"** prompt. You can then discover where your birthday occurs in pi, and if it precedes or is subsequent to some friend's birthday. Happy Pi Searching!

Epilogue

Calculating increasingly more accurate approximations of pi is not difficult with our modern technology, but it is not just the calculation of additional digits that is important. The objective is searching for answers that can explain why something so relatively simple as the ratio of a circle's circumference to its diameter could be involved in such a complex manner throughout mathematics, physics, statistics, engineering, architecture, biology, astronomy, and even the arts. Pi is found in the rhythm of sound waves and ocean waves. It is indeed ubiquitous in all of the Lord's Creation. If this number could be more completely understood, if a pattern could be discovered among its infinite sequence of digits, or if it could be discovered why it is so fundamentally significant in so many apparently unrelated equations – we would then have a better understanding of the mathematics and physics of the cosmos in which we reside as the crowning achievement of the Lord's Creation.

Such is the charm and the mystery of pi – a paradox of simplicity and complexity – from the simple circle to the apparent randomness of its digits. This intriguing number we know as π is irrational because it cannot be expressed as the ratio of two integers, but it is more than irrational; it is transcendental also, because it cannot be expressed as an algebraic equation with a finite number of terms having rational coefficients – but that is another story for another time.

Meanwhile, only the Living Lord God, the designer of all reality, understands the significance and relationships involved with π – and perhaps we may discover the answers, and perhaps we will not know until He reveals them to all who will be in His eternal fellowship. I will be in His Heaven because I have trusted and received Jesus as my personal Savior and Lord. Because I am one who has been blessed by His Love, Mercy, and Grace, I am heir to everything He is and has. Praise Jesus. Why do I think that all believers will share the truth about π and everything else? I believe it is by the authority of God's Word.

According to His Word,
"In the Lord Jesus Christ are hidden all the treasures of wisdom and knowledge."(Colossians 2:3)

AND

"…of making many books there is no end; and much study is a weariness of the flesh. Let us hear the conclusion of the whole matter: Respect and Revere God, and keep His commandments: for this is the whole duty of man. For God shall bring every work into judgment, with every secret thing, whether it be good, or whether it be evil." (Ecclesiastes 12:12-14)

Jesus is Lord. Hallelujah!

Merle

Appendix

Answers for Chapters 14 and 15

Chapter 14 Answer

Elusive Equality

The arrows represent a Roman Numeral equation.
[π (pi) is approximately equal to 22 divided by 7].

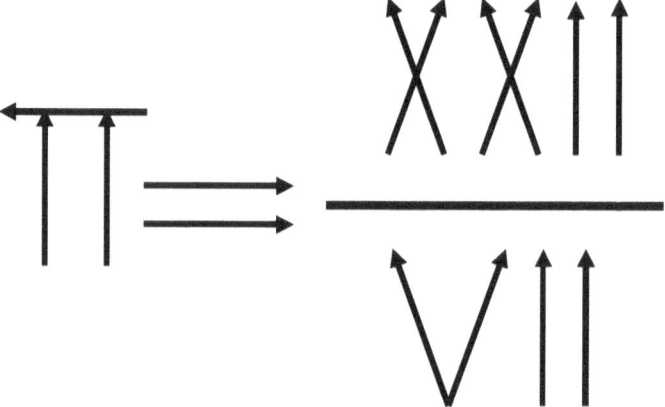

Chapter 15 Answers

(9) **The Lord Jesus**

(11) Pumpkin Pi

Backispiece

A View of the Last Digit of π

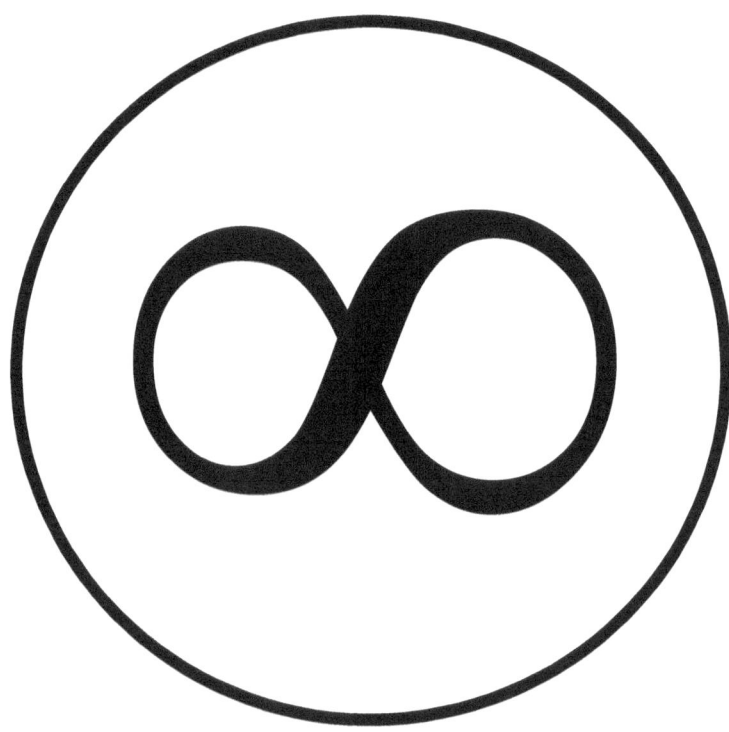

Special Note:
For your viewing pleasure, Appendix Pi, at the back of this book, presents the first 2000 decimal digits of pi.

PART 2

GRAVITATION

GRAVITATION

Caveat Lector

(Caution to the Reader)

Minimally, regardless of your educational status or mathematical proclivity, you can read, without any difficulty, the following sections of this article:

In Chapter 4, "Examples of Gravitational Force" (prior to Riddle). the complete mathematical analysis of six examples is presented. If you feel that the calculations of this chapter may frighten or bore your sensibilities, you can direct your attention to Table 1 (conclusion of chapter) that summarizes the results of all the examples.

Prologue

What is the mysterious force that gives stability to the universe? What is this force that continues to puzzle scientists? How does this force exert its influence across vast distances of space, and more fundamentally, why does it exist? Science is inadequate to explain the force of gravitation as well as all the "natural" laws that God has designed into His universe. These universal laws do not occur by random chance or natural selection. Gravitation, just like all the other physical laws, is a testimony to its Creator.

1
Introduction

You may have heard the story of how a young Isaac Newton began to think about this force after he was hit on the head by a falling apple (Reference the Book cover to view the apple that fell on Isaac). His study led to his discovery of the Universal Law of Gravitation. <u>All</u> objects in the universe attract each other. Specifically, any two bodies attract each other with a force directly proportional to the product of their masses, and inversely proportional to the square of the distance between them. This relationship is expressed by

$$F \propto \frac{M_1 M_2}{d^2}$$

As illustrated in Diagram 1, F is the force of attraction, M_1 and M_2 are the two masses, and d is the distance between them. Mass M_1 attracts M_2 with a force F to the left, and mass M_2 attracts M_1 with an equal force to the right.

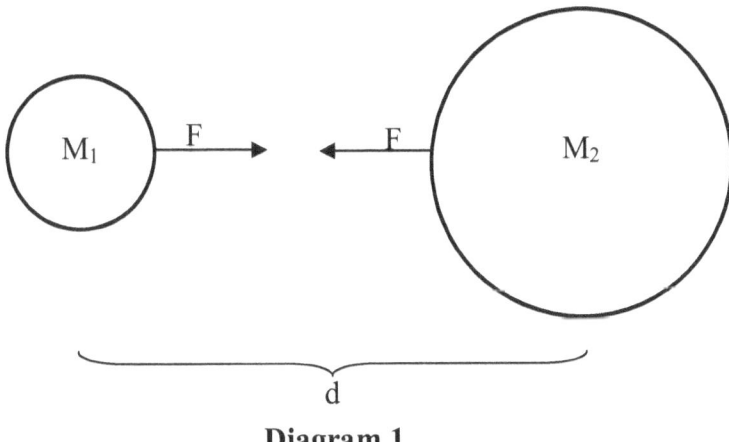

Diagram 1

To make an equation of this relationship, it is only necessary to replace the proportionality symbol with an equal sign and insert G, the "Newtonian Constant of Gravitation".

$$F = G \frac{M_1 M_2}{d^2}$$

2
Properties

The gravitational force has many interesting properties. Only a few will be briefly mentioned here to describe the uniqueness of this force.

G

This force does not change with time – it appears to be an invariable constant.

G

From the perspective of the earth's surface, and neglecting air resistance, all objects, regardless of size, take the same time to fall downward.

G

Gravitation is always an attractive force, unlike magnetism which can also repel.

G

Gravitational force cannot be turned off or negated.

G

This force is dependent only on the masses of the objects. If several objects have equal masses, they will attract each other with identical forces, regardless of the size or composition of the objects.

G

The gravitational force decreases with distance, but its range is infinite. This force exists across millions of light years of space between galaxies as well as between the earth and moon.

3
Fundamental Forces

I think it is singularly fascinating that only <u>four</u> fundamental forces have been discovered in our physical cosmos. Obviously, we are all aware of the effects of many different forces such as tides, wind, electric motors, volcanic and chemical explosions, and human muscles. All of these different kinds of forces, however, are versions of the four basic forces of the universe. Two of these forces, the strong nuclear force, and the weak nuclear force operate entirely within atomic nuclei. Therefore, we are not normally aware of them. The other two forces, electromagnetism and gravitation, account for almost all the forces that we encounter in our everyday lives.

The most powerful of the four forces is the strong force. It binds protons and neutrons together in the nuclei of atoms. It is the force that powers the sun. Although strong, this force is very short-ranged (only within the nucleus of the atom).

The next most powerful force is electromagnetism. This force acts between any particles carrying an electric charge. It keeps negatively charged electrons close to the positive nuclei of atoms throughout chemical reactions. This is the force that enables TV, the telephone, computers, radio, radar, microwaves, light bulbs, and dishwashers. This force is long-ranged, but much weaker than the strong force.

The third most powerful of the four forces is the weak nuclear force. Like the strong force, it also applies only at the subatomic level. The weak force is responsible for radioactive decay processes such as the emission of beta particles from certain nuclei. The weak force is harnessed in modern hospitals in the form of radioactive tracers used in nuclear medicine.

The weakest of the four forces is gravitation. Gravitation is the attraction between any two objects because of their masses. It is the force that keeps the planets in orbit around the sun. Although significantly weaker than the other three forces, the gravitational force acts over great distances.

Assume (for the purpose of approximate relative strength comparisons) that the strongest of the four forces, the strong nuclear force, has a strength index of 1. The relative strengths of the four fundamental forces reveal the feeble force of gravity:

Strong Nuclear Force: 1
Electromagnetic Force: 10^{-2}
Weak Nuclear Force: 10^{-6}
Gravitational Force: 10^{-40}

In other words, the electromagnetic force is 100 times weaker than the strong force, the weak force is a million times weaker than the strong force, and gravitation is (1 followed by 40 zeros) times weaker than the strong force! The relative "weakness" of gravity is singularly interesting, especially when you consider demonstrating it by jumping off the top of a building.

The Physical and the Spiritual are not mutually exclusive; the Physical cosmos and all of its processes are merely subsets of all Reality, which is Spiritual in nature. God's Word is explicit with information regarding His sustaining power for His Creation. In Colossians 1:17, we are told that He (Jesus) existed before all things, and because of His power, all things are held together. In Hebrews 1:3, we learn that the Son (Jesus) is the radiance of God's Glory, and the exact representation of His Being, controlling and sustaining all things by His powerful word.

4
Examples of Gravitational Force

The U.S. Customary System of units will be used for all the examples in this presentation. Reference **Appendix 1** for the conversion of the gravitational constant from the mks System to the U.S. Customary System.

For simplification, the following examples will not indicate all the units of measurement for values of the various variables. For those who may be interested in details of the <u>complete</u> dimensional analysis of the variables in the gravitation equation, refer to **Appendix 2**. I have developed the dimensional analysis for the gravitation equation to emphasize that it is just as important to have the correct relationship among the various units as it is to have the correct numerical values. Numerical calculations will be correct <u>only</u> if the units of measurement associated with each of the calculations are also correctly related mathematically.

Reference **Appendix 3** for computational data used in the following examples.

Example 1: You and the Moon

What is the gravitational force between you and the moon?

$G \approx 3.439 \times 10^{-8}$

M_1 (Mass of Moon)

$$\approx 7.35 \times 10^{22} \, kg \approx 7.35 \times 10^{22} \, kg \times 2.204622341 \frac{lb}{kg}$$

$$\approx 1.620397421 \times 10^{23} \, lb$$

$$\frac{1.620397421 \times 10^{23} \, lb}{32.2 \, ft/s^2} \approx 5.032290127 \times 10^{21} \, slugs$$

$$M_2 \, (\text{Your Mass: Assume 150 lb}) \approx \frac{150 \, lb}{32.2 \, ft/s^2} \approx 4.658 \, slugs$$

Distance from Earth to Moon $\approx 3.84 \times 10^8 \, m$

$\quad 3.84 \times 10^8 \, m \times 3.280833 \, ft/m \approx 1,260,000,000 \, ft$

Earth Radius $\approx 3960 \, mi$

$$3960 \ mi \times 5280 \ ft/mi \approx 20,908,800 \ ft$$

Therefore $d \approx 1,260,000,000 - 20,908,800 \approx 1,239,091,200 \ ft$

$$F = G\frac{M_1 M_2}{d^2}$$

$$F \approx 3.439 \times 10^{-8} \times \frac{\left(5.03 \times 10^{21}\right)\left(4.658\right)}{\left(1,239,091,200\right)^2} lb$$

$$\approx 5.24799122 \times 10^{-4} \ lb \approx .0005 \ lb \left(.008 \ ounce\right)$$

Example 2: Two Locomotives

Two locomotives, each weighing 64 tons, are positioned side by side with their centers 10 ft. apart. What is the gravitational force between them?

$$G \approx 3.439 \times 10^{-8}$$

M_1 and M_2 (Mass of locomotives)

$$\approx \frac{64 \times 2000}{32.2} \approx 3975.155 \ slugs$$

$$d = 10$$

$$F = G\frac{M_1 M_2}{d^2}$$

$$F \approx 3.439 \times 10^{-8} \times \frac{\left(3975\right)\left(3975\right)}{10^2} \approx .005434 \approx .005 \ lb \left(.08 \ ounce\right)$$

Example 3: You and the Earth

What is the gravitational force between you and the earth?

$$G \approx 3.439 \times 10^{-8}$$

$$M_E \approx \frac{5.98 \times 10^{24} \ kg \times 2.2046 \dfrac{lb}{kg}}{32.2} \approx 4.09 \times 10^{23} \ slugs$$

Assume you weigh 150 lb:

$$M_{(you)} \approx \frac{150 \ lb}{32.2} \approx 4.658 \ slugs$$

$$d \approx 3960 \ mi \times 5280\frac{ft}{mi} \approx 20,900,000 \ ft$$

$$F = G\frac{M_E M_{(you)}}{d^2}$$

$$F \approx 3.439 \times 10^{-8} \times \frac{\left(4.09 \times 10^{23}\right)\left(4.658\right)}{\left(20,900,000\right)^2}$$

$$F \approx 149.99 \; lb$$

Example 4: Earth and Moon

What is the gravitational force between the earth and the moon?

$G \approx 3.439 \times 10^{-8}$

$$M_E \approx \frac{5.98 \times 10^{24}\,kg \times 2.2046\,\dfrac{lb}{kg}}{32.2} \approx 4.09 \times 10^{23}\,slugs$$

$$M_M \approx \frac{7.35 \times 10^{22}\,kg \times 2.2046\,\dfrac{lb}{kg}}{32.2} \approx 5.03 \times 10^{21}\,slugs$$

Distance (d) between earth and moon:

$$d \approx 3.84 \times 10^{8}\,m \times 3.2808\,\frac{ft}{m} \approx 1,259,827,200\,ft$$

$$F = G\frac{M_E M_M}{d^2}$$

$$F \approx 3.439 \times 10^{-8} \times \frac{\left(4.09 \times 10^{23}\right)\left(5.03 \times 10^{21}\right)}{\left(1.26 \times 10^{9}\right)^2}$$

$F \approx 4.5 \times 10^{19}\,lb$ (or about 22.5 million billion tons)

Example 5: Earth and Sun

What is the gravitational force between the earth and the sun?

$G \approx 3.439 \times 10^{-8}$

$M_E \approx 4.09 \times 10^{23}\,slugs$ (reference calculation in Example 4)

$$M_S \approx \frac{1.99 \times 10^{30}\,kg \times 2.2046\,\dfrac{lb}{kg}}{32.2} \approx 1.36 \times 10^{29}\,slugs$$

Distance (d) between earth and sun:

$$d \approx 1.50 \times 10^{11}\,m \times 3.2808\,\frac{ft}{m} \approx 4.92 \times 10^{11}\,ft$$

$$F = G\frac{M_E M_S}{d^2}$$

$$F \approx 3.439 \times 10^{-8} \times \frac{\left(4.09 \times 10^{23}\right)\left(1.36 \times 10^{29}\right)}{\left(4.92 \times 10^{11}\right)^2}$$

$F \approx 8 \times 10^{21}\,lb$ (or about 4 billion billion tons)

Example 6: Two Persons (Two Feet Apart)

What is the gravitational force between two persons that are two feet apart?

$G \approx 3.439 \times 10^{-8}$

Weight of Person 1 = 150 lb
Weight of Person 2 = 100 lb

$$M_{P1} \approx \frac{150 \ lb}{32.2 \ ft/s^2} \approx 4.658 \ slugs$$

$$M_{P2} \approx \frac{100 \ lb}{32.2 \ ft/s^2} \approx 3.106 \ slugs$$

Distance (d) between the two persons = 2 ft (distance between center of masses)

$$F = G \frac{M_{P1} M_{P2}}{d^2}$$

$$F \approx 3.439 \times 10^{-8} \times \frac{(4.658)(3.106)}{(2)^2}$$

$$F \approx 1.244 \times 10^{-7} \ lb \approx 1.99 \times 10^{-6} \ ounces$$
$$\approx 2 \ \text{millionths of one ounce}$$

$$\left(\frac{2}{1,000,000} \ ounce \right)$$

If these two persons moved so that they are 6 ft apart, the attractive force is 22 one-hundred millionths of one ounce $\left(\frac{22}{100,000,000} \ ounce \right)$ because they are now 3 times further apart, and the force varies inversely with the square of the distance – the force is 9 times smaller. This can be verified by modifying the above calculation for F.
You do the math.

Table 1 – Examples of the Gravitational Force Between Objects

Example #	Objects	Gravitational Force (Pounds)
1	You and the Moon	.0005 (.008 ounce)
2	Two Locomotives	.005 (.08 ounce)
3	You and the Earth	(Your Weight)
4	Earth and Moon	4.5×10^{19} (22.5 million billion tons)
5	Earth and Sun	8×10^{21} (4 billion billion tons)
6	Two Persons (2 Feet Apart)	.0000001244 (.000002 ounce) (2 millionths of one ounce)

5
An Alphabetic Anthropomorphic Riddle

I am given Divine
       Authorization

To cause universal
       Benefaction.

To each and to every
       Civilization,

I have always shown my
       Dedication.

I often create some
       Exclamation,

And I sometimes will cause
       Frustration.

I act with absolutely no
       Gradation,

And everywhere I make my
       Habitation,

Everything follows my
       Inclination.

In Nepal there is real
       Justification

That I'm less strict than in
       Kingston.

My nature permits no
        Levitation,

But you may see a fake
        Manifestation.

My clever method of
        Navigation

Has not been proven by
        Observation.

My skill is performed with
        Perfection,

And this is true without
        Qualification.

I have a long-standing
        Reputation –

I'm consistent in each
        Situation.

I'm never guilty of
        Transgression,

And I'm in favor of
        Unification.

Silently and without
        Vexation,

I am present at any
        Workstation.

I'm at home in space with
        X-radiation,

Or on the sea with a
      Yachtsman.

My influence has no
      Zonation.

WHO AM I?

Epilogue

In addition to the forces and laws inherent in our physical universe, there are also, of course, civil and criminal laws. Consider the multiple millions of laws that are enacted throughout the world by various political, legal, and religious institutions (in efforts to merely enforce the original ten). Then consider the prevalence of the perversion of these laws by immoral practices for mercenary motives without regard to what is right, honorable, or just. In instance after instance, individually and corporately, in our nation and internationally, dishonorable actions destroy the integrity of any system that attempts to promote the principles of ethical and moral conduct. This moral pollution reflects the decadence of unregenerate human nature. Consequently, the applications of laws are contaminated by partiality, dishonesty, and procrastinated action.

However, praise the Lord Jesus who can and does save us from our sinful nature (if we choose, by faith, to receive His salvation). He has established laws in His Spiritual realm and designed laws in His physical cosmos through which He has given us life and purpose. Truth, justice, righteousness, and judgement will prevail according to His plan and time.

Yes, His plan will prevail, and is proceeding on schedule. The ultimate reality is that:

> On some great day with one accord,
> Each knee will bow and tongue confess
> That Jesus Christ is Lord.

Meanwhile, I conclude this article by asking you a question. Contrary to the depravity that is pervasive among us, there is a law operative in our world that is administered without prejudice, without corruption, and without delay. This law has a profound and continual affect on your life. Do you know this law?

Reference the next page for the answer.

Answer:

The Universal Law of Gravitation
(This is <u>also</u> the answer to the Riddle)

Appendix 1

Gravitational Constant (G)

$$G(mks) \approx 6.672 \times 10^{-11} N \bullet \frac{m^2}{kg^2}, \text{ where}$$

$$1N(Newton) = 1kg \bullet \frac{1m}{s^2}$$

Therefore, $G(mks) \approx 6.672 \times 10^{-11} \dfrac{m^3}{kg \ s^2}$

Converting G from the mks System to the U.S. Customary System:

$$6.672 \times 10^{-11} \frac{m^3}{kg \ s^2} \times \frac{35.314 \ ft^3}{m^3} \times \frac{14.594 \ kg}{slug}$$

$$\approx 3.438565427 \times 10^{-8} \frac{ft^3}{slug \ s^2}$$

$$G(\text{U.S. Customary}) \approx 3.439 \times 10^{-8} \frac{ft^3}{slug \ s^2}, \text{ where the slug} =$$

$$\frac{lb}{ft/s^2}$$

Note:
"mks" is an abbreviation for meter, kilogram, second.
These metric units are part of the International System of Units (SI).
The U.S. Customary System (also known as the Conventional System) is derived from the traditional Imperial System of the United Kingdom.

Appendix 2

Dimensional Analysis for Force of Gravitation

$$F = G\frac{M_1 M_2}{d^2}$$

Dimensions:

$$\text{Gravitational Constant } (G) = \frac{ft^3}{\left(\dfrac{lb}{ft/s^2}\right)(s^2)}$$

[Reference Appendix 1]

$$\text{Mass } (M_1 \text{ and } M_2) \text{ in slugs} = \frac{lb}{ft/s^2}$$

Distance (d) = ft

Force (F) = lb

$$F = \left[\frac{ft^3}{\left(\dfrac{lb}{ft/s^2}\right)(s^2)}\right]\left[\frac{\left(\dfrac{lb}{ft/s^2}\right)\left(\dfrac{lb}{ft/s^2}\right)}{ft^2}\right]$$

$$G = \frac{ft^3}{\left(\dfrac{lb\,s^2}{ft}\right)s^2} = \frac{\dfrac{ft^3}{lb\,s^4}}{ft} = \frac{ft^3\,ft}{lb\,s^4} = \frac{ft^4}{lb\,s^4}$$

$$\frac{M_1 M_2}{d^2} = \frac{lb^2}{\dfrac{ft^2/s^4}{ft^4}} = \frac{lb^2\,s^4}{\dfrac{ft^2}{ft^2}} = \frac{lb^2\,s^4}{ft^4}$$

$$F = \left(\frac{ft^4}{lb\,s^4}\right)\left(\frac{lb^2\,s^4}{ft^4}\right) = lb \text{ (Gravitational Force in pounds)}$$

Appendix 3

Computational Data

Earth Data

          Mean Radius ≈ 3960 miles (3958.7 mi)

      Distance from sun (mean) $\approx 1.50\times10^{11}\, m\,(9.30\times10^7\, mi\,)$

$$\approx 4.92125\times10^{11}\, ft$$

      Distance from moon (mean) $\approx 3.84\times10^8\, m\,(2.39\times10^5\, mi\,)$

      Mass $\approx 5.98\times10^{24}\, kg$

Mass of Sun $\approx 1.99\times10^{30}\, kg$

Mass of Moon $\approx 7.35\times10^{22}\, kg$

$g \approx 32.2\, ft/s^2$ or $9.81\, m/s^2$ at earth's surface

Gravitational Constant (mks System) $\approx 6.672\times10^{-11}\, N\bullet\dfrac{m^2}{kg^2}$

$$\approx 6.672\times10^{-11}\,\dfrac{m^3}{kg\ s^2}$$

Gravitational Constant (U.S. Customary System)

$$\approx 3.439\times10^{-8}\,\dfrac{ft^3}{slug\ s^2}$$

$m \approx 3.280833\, ft$

$kg \approx 2.204622341\, lb$

mi = mile; m = meter; ft = feet; kg = kilogram

g = acceleration of gravity; s = second; lb = pound

$1\text{N (Newton)} = 1kg\bullet\dfrac{1m}{s^2}\approx .2248\, lb\,;\, 1\, lb \approx 4.448 N$

$1\text{ slug} = \dfrac{1lb}{1\, ft/s^2} = 1lb\bullet\dfrac{s^2}{ft}\approx 32.2\, lb$

Appendix Pi

On the following pages you will observe the first 2000 decimal digits of pi. The beginning of this approximation of pi is the only integer and decimal point:

3.

π

1415926535

8979323846

2643383279

5028841971

6939937510

π

5820974944

5923078164

0628620899

8628034825

3421170679

π

8214808651

3282306647

0938446095

5058223172

5359408128

π

4811174502

8410270193

8521105559

6446229489

5493038196

π

4428810975

6659334461

2847564823

3786783165

2712019091

π

4564856692

3460348610

4543266482

1339360726

0249141273

π

7245870066

0631558817

4881520920

9628292540

9171536436

$$\pi$$

7892590360

0113305305

4882046652

1384146951

9415116094

π

3305727036

5759591953

0921861173

8193261179

3105118548

π

0744623799

6274956735

1885752724

8912279381

8301194912

π

9833673362

4406566430

8602139494

6395224737

1907021798

π

6094370277

0539217176

2931767523

8467481846

7669405132

π

0005681271

4526356082

7785771342

7577896091

7363717872

π

1468440901

2249534301

4654958537

1050792279

6892589235

π

4201995611

2129021960

8640344181

5981362977

4771309960

π

5187072113

4999999837

2978049951

0597317328

1609631859

$$\pi$$

5024459455

3469083026

4252230825

3344685035

2610311881

π

710100313

7838752886

5875332083

8142061717

7669147303

π

59825343904

2875546873

1159562863

8823537875

9375195778

π

1857780532

1712268066

1300192787

6611195909

2164201989

π

3809525720

1065485863

2788659361

5338182796

8230301952

π

0353018529

6899577362

2599413891

2497217752

8347913151

$$\pi$$

5574857242

4541506959

5082953311

6861727855

8890750983

π

8175463746

4939319255

0604009277

0167113900

9848824012

$$\pi$$

8583616035

6370766010

4710181942

9555961989

4676783744

π

9448255379

7747268471

0404753464

6208046684

2590694912

π

9331367702

8989152104

7521620569

6602405803

8150193511

π

2533824300

3558764024

7496473263

9141992726

0426992279

π

6782354781

6360093417

2164121992

4586315030

2861829745

π

5570674983

8505494588

5869269956

9092721079

7509302955

π

3211653449

8720275596

0236480665

4991198818

3479775356

$$\pi$$

6369807426

5425278625

5181841757

4672890977

7727938000

π

8164706001

6145249192

1732172147

7235014144

1973568548

π

1613611573

5255213347

5741849468

4385233239

0739414333

π

4547762416

8625189835

6948556209

9219222184

2725502542

π

5688767179

0494601653

4668049886

2723279178

6085784383

π

8279679766

8145410095

3883786360

9506800642

2512520511

$$\pi$$

7392984896

0841284886

2694560424

1965285022

2106611863

π

0674427862

2039194945

0471237137

8696095636

4371917287

π

4677646575

7396241389

0865832645

9958133904

7802759009

These 2001 digits plus an infinitude of additional digits comprise the value of pi...